WATER SCARCITY SOLUTION AND AI BASED FUTURE MANAGEMENT

Arman Mohammad Nakib

Title: **WATER SCARCITY SOLUTION AND AI BASED FUTURE MANAGEMENT**

ISBN: 979-8-89248-584-5

Author: Arman Mohammad Nakib

Cover image:
https://pixabay.com/ru/images/search/artificial%20intelligence/?pagi=13

Publisher: Generis Publishing
Online orders: www.generis-publishing.com
Contact email: info@generis-publishing.com

Water Scarcity Solution

and

AI Based Future Management

Arman Mohammad Nakib

Special thanks to

Professor Dr. Yuemei Luo (MSc Supervisor),
Jobaydul Hasan Emon (Friend),
Sakib Chowdhury (Friend),
Badhon Barua (Student)

ABSTRACT

Wastage of water is a burning topic in the world. Different countries worldwide are facing the issue of the lack of fresh water, and the problem is increasing daily. This book is based on a paper that aims to design a system that will predict the amount of water needed by a family or in a locality depending on family members, region, temperature, season, occupation, location, and religion. It can also be possible to forecast the water demand in a locality, area, or country depending on these factors. Previous works in this field focus on something other than these factors and the distribution system mentioned in this paper. Different machine learning models will predict the amount of water required by a family or a locality based on these factors. Then, water will be supplied using these expected values so that each family or community in a locality receives the desired amount of water. The practical circuit uses an Arduino microcontroller, water flow meter, solenoids, etc. Water distribution is automatically controlled by the water flow meter and solenoid, and no family or community in a locality will receive more water than the predicted values per day. So it will reduce water wastage, and everybody will use it according to their daily needs. Different machine learning models were used in this proposed design to compare the performance of the models for this task. Linear, Ridge, Lasso, ElasticNet, Decision Tree, Random Forest, XGBoost (Extreme Gradient Boosting), KNN (K-Nearest Neighbors), SVR (Support Vector Regression), MLP (Multilayer Perceptron), LightGBM (Light Gradient-Boosting Machine), CatBoost, Deep Neural Network have been used. Different model's performances have been analyzed. The analyzing factors are model training time, model prediction time, Robustness to outliers, and scalability. All these performances were analyzed to determine which model is best for this work. So, the Decision Tree and LightGBM models are the best based on comparing all the models for this task.

Contents

Chapter 1.
INTRODUCTION

The world is dealing with many problems due to climate change, and those are mainly for the limited resources to live with. Most of the problems are created by humans, and their lifestyle, lack of awareness, lack of knowledge, etc, are the common factors that make life difficult on this beautiful earth. Among all the resources, water is a fundamental need for all living beings. However, many countries worldwide are facing the issue of fresh water. An average family can waste around 100 liters (www.epa.gov/watersense/statistics-and-facts) per day due to water leakage, according to the information given by the United States Environmental Protection Agency. However, a person's absolute essential consumption of water per day is about 92 liters (Crouch, M. L., Jacobs, H. E., & Speight, V. L., 2021). So many amounts of water are wasted every day all around the world. Countries like Australia, Russia, Germany, the United Kingdom, Egypt, France, Brazil, Iran, and more wastewater the most.

The amount of water a person uses daily depends on factors like region, season, temperature, location, Occupation, and religion (Almulhim, A. I., &Abubakar, I. R., 2024). These combinations of factors should have been introduced in previous research works. People who live in the Middle East use more water than people in America, Europe, and Australia. In comparison, people in Asia and Africa use much less water than other region countries (Timmerman, M., 2013; Utami, R. R., Geerling, G. W., Salami, I. R., Notodarmojo, S., & Ragas, A. M., 2024; www.statista.com/statistics/278066/global-water-demand-by-region/). Similarly, people use more water in summer rather than in other seasons. That means the countries where summer is the primary season are where people will use more water than in places where summer is not. That means high temperature leads people to use more water than medium or low-temperature situations (Bergel, T., &Młyńska, A., 2021; Dimkić, D., 2020; Rondinel-Oviedo, D. R., & Sarmiento-Pastor, J. M., 2020; Rodríguez, C., Sánchez, R., Lozano-Parra, J., Rebolledo, N., Schneider, N., Serrano, J., &Leiva, E., 2020, Xenochristou, M., &Blokker, M., 2018, July). People in urban areas use more water than people in rural areas (Chang, H., Praskievicz, S., &Parandvash, H., 2014; Fan, L., Liu, G., Wang, F., Geissen, V., &Ritsema, C. J., 2013; Matos, C., Briga-Sá, A., Pereira, S., & Silva-Afonso, A., 2013, September; Yan, L., 2015). Another important thing is that people's Occupation influences water use daily. General laborers use more water than the official job holders, while students and jobless people are moderate water users between the two listed occupations (Singha, B., Eljamal, O., &Karmaker, S. C., 2024).

Religious people in Islam use more water than people of other religions due to their purification works (Smith, A., & Ali, M., 2006). So these are the critical factors that control water use, or the amount of water people use.

So, in this paper, these factors are considered to build up a sample dataset according to the references provided, as no recognized and organized dataset fills up the demand of our requirements. So, data from the different references provided earlier was collected, and a sample dataset was built. So, in the Python code, we generated random data according to the other factors listed earlier and calculated the average amount of water needed by a person per day. Then, data modeling was done. One hot encoding is used to convert categorical variables into a format that can be provided to the algorithm. Data is divided into training, validation, and test data, and machine learning is used to predict the daily water needed for this paper. Different machine learning models and MLP and deep neural network models were used to compare the output results.

The paper is designed to be implemented in a house or big house where many families live together. However, the concept provided in the paper can be easily implemented in big cities or towns or even in the whole country to manage water distribution properly and reduce the waste of time. The nice part of this proposed design is that the user can input data from the outside, such as how many families and family members are in their family, etc. The house owner will give input on the total water in the tank and the number of families, and all the families can provide input on the number of family members from their homes. So, the machine learning model will predict the daily water needed by a person depending on the family members and give input about their region, location, occupation, religion, the season of where they are, the temperature they are facing, and the number of family members they have. The machine learning model predicts the water a person needs per day based on the six factors used in the design. Then, according to the total amount of water in the water tank and the number of families in each family, the Python code will calculate the percentage of total water needed by each family by using the predicted water demand value by a person and the total number of a family member in a family. So, the machine learning model will give the expected value for one person, and the rest of the code will calculate the total water needed by each family independently according to the number of family members. It can also be applied to supply water in a locality.

The noticeable part of the design is that each family will give input from their house accordingly. So it will not create any hassle. The total water in a tank will be distributed to all the families according to their daily demands, considering all the factors and the number of family members. This distribution is not equal. This water is distributed according to the needs of that family.

The last part of the design is the Arduino microcontroller design, which is used for practical implementation. Machine learning and other Python codes provided the total water each family needed. Then, each family water demand value in liters would be given in the Arduino C++ code. In the proposed design, water flow sensors and solenoids are used to control the water flow to each family. The maximum water a family needs on that particular day is provided in the code. Water will be delivered to each family through a water flow sensor, recording the water the family receives. When the maximum value is reached, the solenoid will be closed to stop the water supply to that particular family. This thing will happen simultaneously for each of the family. So, the family will get water according to the machine's prediction. As a result, all the families will try to use the water they need. They will save their water as they will know they will get only specific water per day according to their demand predicted by the machine with their provided data. So, this is the proper water management and an appropriate way to reduce water wastage. This distribution is not even water distribution. Some families may have more family members than another family. So, evenly distributed water distribution is not a proper judgment for distributing water. This proposed design distributes water by considering all the factors and predicting based on what amount a particular family needs on a specific day. Automatic control of the water flow and cutting off the solenoid when maximum water is supplied to a particular family makes this design more sophisticated, as all the processes are automotive. The user must provide input data, and the rest of the work will be done automatically. Yes, some families may need extra specifics for emergencies. For that, an additional water reserve system can be used in emergencies. The process is simulated in the Tinkercad and is practically implemented in the circuit to distribute water in different families' houses. The concept can apply to water demand forecasts for a city or a country's various regions.

Chapter 2.
LITERATURE REVIEW

Many works have been done earlier on a similar topic. Some works are related to even water distribution (Kalyani, C., Pradnya, K., Samkit, C., &Patil, V., 2023; Pratama, M., &Firmansyah, M). Very few works were found to do it by machine learning precisely on this topic. Those papers work on the usage of water by people over a long period and then predict the consumer level (Reddy, D. G., V, D., Salanke, N. G. R., & MN, M., 2024; Tejas, T., Yadav, U., Krishnan, V., Vishrutha, V., &Mallikarjuna, M., 2024, February). As the design in this paper considers the different factors and family members than it predicts, it is a more practical way to do this work. All the other works have made water flow control, distribution, water quality measurement, water usage, water level control, water theft reduction, leakage control, waste control by controlling the water flow, etc. All the maximum works depend on the IoT technology and they are used to control all the things remotely (Babu, P., &Rajasekaran, C., 2016; Ali, A., Rahman, M., Siddique, M., Galib, S., &Nashiry, A., 2023; Adu-Manu, K. S., Adjetey, C., &Apea, N. Y. O., 2022; Ali, S., Gillani, S. M. D., Khan, S. A., &Zain, S. M., 2024; Baranidharan, S., Hariprasad, T., Nanthakumar, C., & Kumar, C., 2023, September; Chavhan, S., Gawande, U. H., Bhoyar, D., Rangari, S., Gardi, M., &Gaikwad, A., 2024; Gore, S., Dutt, I., Dahake, R. P., Khodke, H. E., Kurkute, S. L., Dange, B. J., & Gore, S., 2023; Kumar, B. S., Ramalingam, S., Balamurugan, S., Soumiya, S., &Yogeswari, S., 2022, October; Mule, A., Raj, R., Kumar, S., &Rohile; Nirmala, D., Pooja, G., Sowmya, U., Mohamed, A., Durdona, A., Rajavarman, R., &Parkunam, N., 2023; Nur'Im, J., &Setyawan, G., 2023, January; Ogidan, O. K., Olla, M., &Odey, I., 2022, April; Rexline, S. J., Renold, M., &Deeba, M., 2023, October; Rahmatulloh, A., Supriatna, G. T., Widiyasono, N., &Darmawan, I., 2023; Vladislav, D. S., 2022; Velayudhan, N. K., Pradeep, P., Rao, S. N., Devidas, A. R., & Ramesh, M. V., 2022). These were excellent research works as they can control water use remotely, and all are IoT devices. Also, much work has been done on quality control of the water supplied. This paper tried to solve the shortcomings and gaps by considering all the work that has been done before. Even water distribution to all is a good concept, but even water distribution is only one of the solutions. Because all the families in a big house or a city will require different amounts of water, it is not a justice to give even water to all. So, this paper solves the problem of distributing water according to the family's needs. Other works also focused on something other than the automatic cutting of the water supply according to a particular family's water demand by predicted machine learning valuing. So, in this paper, it is done, and it is an exact solution to reduce water waste. None of the research works before didn't work on the factors that influence the amount of water a family will use, such as occupation, religion, region, location, season, etc. This paper also considers predicting the water needed by each family in a locality using different machine-learning models. It has compared the results and provided the best model. The owner and each family can give input on the necessary data, which makes this work unique from the existing works. So, it can be easily applicable to any city or even the whole country to forecast water demand regularly.

Chapter 3.
METHODOLOGY

3.1 Block Diagram: Now we will describe in details about the methodology of the whole proposed design. Figure 1 shows the proposed method of this paper sequentially.

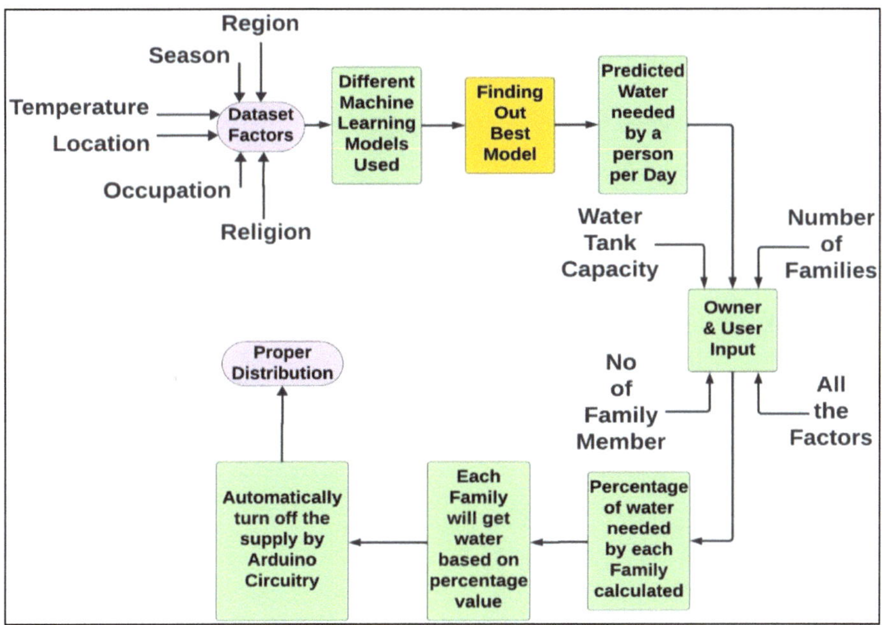

Figure 1: Block diagram of the whole process

Dataset Factors:

It also contains selections for the input of factors such a geographical environment, season, temperature, geographical location, occupation, and religion respectively. All these factors contribute in the formation of the dataset which in turn provides the basis on which machine learning analysis is done.

Different Machine Learning Models:

This segment of the system entails extraction of several factors from the dataset together with the use of a number of machine learning models in the attempt to predict daily water demand. We used several machine learning models, these are Deep Neural

Network, CatBoost Regressor, LightGBM Regressor, MLP Regressor, SVR, KNN, XGBoost, Random Forest, Decision Tree, ElasticNet Regression, Lasso Regression, Ridge Regression, Linear Regression.

Finding Out Best Model:

Having gone through different models the system evaluates the viability of each in an effort to come up with the right model needed for the prediction of water needs. It has been found that Decision Tree and LightGBM models are the best models.

Predicted Water needed by a Person per Day:

Given this input data, learning model will help in giving an estimate of water requirements within the day and the people involved, this will be friendly to planning the flow of water.

Owner & User Input:

This block seeks additional information from the owner and the users of the water system for such aspects like the number of people in a family is likely to improve the distribution ratios.

Owner will give the input of the followings:

a) Water Tank Capacity:

Owner will give input of water tank capacity and this will further be distributed according to consider by other factors.

b) Number of Families:

Owner will also give the input of the number of families that use the water supply help us to will ration the water in the proportion as to each family usage.

c) Others:

Owner will give input like season and temperature (also can be given by users).

User will give the input of the followings:

a) No of Family Member:

This input provides the total population of the people in each family which is vital in partitioning the available water in proportion to families' people.

b) Others:

User will give others input like region, location, occupation and religion.

Percentage of Water needed by Each Family Calculated:

Deriving from the expected individual water consumption and numbers of families, it determines the precise part of total water that should be provided to each of them.

Each Family will get Water based on Percentage Value:

The water is therefore distributed to the families in the ration that was calculated based on the need percentage as cooked up by the model.

Automatically Turn Off the Supply by Arduino Circuitry:

Since the families' water ration is supposed to be rationed and given in accordance with the plan above, an automated mechanism which uses Arduino circuitry shuts off the water faucet after a given family's portion of water is released.

Proper Distribution:

Last of this, it ensures that the Water distribution runs as planned in the planning phase and controlled in delivery so that it is used rightly and suitably by the members of the community.

Now we will describe different sections of the methodology more in details.

3.2 Dataset Preparation

One of the unique parts of this paper is predicting the water needed by a family depending on some critical factors by which the dataset is prepared. North America, Europe, Asia, Africa, the Middle East, Australia, and South America are the subsections of the region. Winter, spring, summer, and fall are the subsections of season. Hot, moderate, and cold are the subsections of temperature. Urban and rural are the subsections of location. Office, labor, student, and jobless are the subsections of Occupation. Islam, Christianity, Hinduism, Buddhism, and others are the subsections of religion. Different water usage values have been taken from different references for a person per day in liters.

Then, the average value of water used per day has been calculated for a particular person. Weights are assigned to all the subcategories of all factors to prepare the dataset.

$$\textbf{Weight} = \frac{\textbf{Water usage value per day of a subcategory}}{\textbf{Total water usage value of all subcategories in a factor}}$$

The above equation calculates weights. Then, random data is generated to create a data frame. That is how the sample dataset was created. A 1000-row combination was generated for the dataset (shown in Table 1).

We will give an example that how it was calculated.

Let we take Temperature factor from our dataset. The categories under temperature and average daily water usage values are:

Hot: 256 liters per day

Moderate: 161 liters per day

Cold: 114 liters per day

The total water usage value is:

256 + 161 + 114 = 531 liters per day

Now below, calculation of finding out weight for each subcategory is shown:

Weight (Hot) = $\frac{256}{531}$ = 0.482

Weight (Moderate) = $\frac{161}{531}$ = 0.303

Weight (Cold) = $\frac{114}{531}$ = 0.215

Weights also give us indication of which factors are more important than the others. That is how all the subcategories of all the factors are calculated. Suppose, we generate one row randomly like Asia, Summer, Moderate, Urban, Student, Hinduism. This is the one row of the table. As we know the all weights and the all water usage values of all subcategories, we will now calculate the water needed for a person from Asia, in summer, temperature is moderate, the person is from urban area and he or she is a student and follows Hinduism. The calculation is:

Daily water usage by a person = (132 × 0.062) + (256 × 0.369) + (161 × 0.303) + (284 × 0.536) + (227 × 0.211) + (284 × 0.195) = 406.932 liters per day

Table 1. *Random 10 Rows from Dataset*

Sl No	Region	Season	Temperature	Location	Occupation	Religion	Daily Water Usage
0	North America	spring	moderate	urban	office	buddhism	423.047
1	Asia	spring	hot	urban	student	christianity	424.985
2	Middle East	summer	moderate	rural	labor	christianity	613.649
3	Europe	fall	moderate	rural	jobless	others	380.867
4	South America	summer	moderate	urban	labor	islam	494.609
5	Middle East	winter	hot	urban	jobless	others	611.114
6	Middle East	winter	cold	rural	student	hinduism	465.961
7	Asia	summer	moderate	urban	student	islam	422.172
8	Australia	summer	hot	urban	office	buddhism	543.178
9	South America	summer	hot	rural	labor	christianity	515.898

Dataset is the most critical part of a research. We had to pass much time to prepare an at least a decent dataset. We have to collect values from different resources. We didn't find any build in dataset based on what dataset we need. So we had to build it by our own way. Factors we added are actually influence the water usage on a person in daily works and the table below is randomly generated different combination of the factors of a person from which area and how he or she is in his or her practical life. This makes this research strong to predict or forecast of water needed by a family or in an area. Table shows the 10 random rows to understand the data we used here. Last row of the table shows the daily water usage of a person which can be used to calculate the total family needs also.

3.3 Machine Learning Models

Different machine learning models have been used to predict the water needed by a person per day depending on the region, season, temperature, location, Occupation, religious factors, and their subsections. Linear, Ridge, Lasso, ElasticNet, Decision Tree, Random Forest, XGBoost, LightGBM, CatBoost, SVR, KNN, MLP, and Deep Neural Network models are used to compare the results of the output predictions. Different regression models are used along with Multilayer Perceptron (MLP), a modern feedforward Artificial Neural Network (ANN). Deep Neural Network is also used to see and compare the output predictions.

Now we will give the explanation of the machine learning model used in our proposed design.

a) Linear regression is a linear approach for modeling the relationship between a dependent variable and one or more independent variables.

$$y = \beta_0 + \beta_1 x_1 + \beta_2 x_2 \text{ ------- } + \beta_n x_n + \text{error}$$

β_0 = a constant

β_1, β_2, ---, n = regression coefficients

y = dependent variable

x_1, x_2, ---, n = independent variables

b) Ridge regression is also a linear regression approach but it includes a regularization term. It uses a penalty to the sum of the squares of the model coefficients. In the code, we used alpha = 1.0 which is the regularization (L2) strength. The larger the value, the stronger the regularization is.

$$\text{Min. obj.} = \text{Least Squares Objective (no regularization)} + \alpha * ||\beta||^2$$

Min. obj. = Minimization Objective

β is the weights of the coefficients (independent variables)

α is a regularization parameter (for strength of the penalty term)

c) Lasso abbreviated as Least Absolute Shrinkage and Selection Operator. This regression adds a regularization (L1) term to the cost function for the absolute values of the coefficients. If some coefficients become zero, it gives the method of feature selection.

$$\text{Min. obj.} = \text{Least Squares Objective (no regularization)} + \alpha * |\beta|$$

In the code, alpha = 0.1 is used.

d) If we want to combine both the properties of Ridge and Lasso regression, then we need to use ElasticNet. It uses both L1 norm and L2 norm.

$$\text{Min. obj.} = \text{Least Squares Objective (no regularization)} + \alpha * |\beta| + \alpha * ||\beta||^2$$

The parameter we used in the code is:

alpha = 0.01 (scales the penalty)

l1_ratio = 0.1 (used to regularization between L1 and L2)

e) Decision tree regression is a non-linear model which is based on tree structure. It splits the dataset into smaller subsets by considering the feature values and it minimizes cost function very similar to the mean squared error for every node. At each node, it finds out the best division. We used 'random_state = 42' in the code that ensures the result of the tree structure (algorithm) is always same for every run of the code with the same data. We can use any number but 42 is a random choice.

f) Random Forest gives the prediction by averaging the predictions of every individual tree. Random Forest improves the model accuracy and also reduces over-fitting. In our case, 'n_estimators = 100' is set as parameter which is the total number of trees in the forest. 'random_state = 42' is set again.

g) Extreme Gradient Boosting is the full form of XGBoost. It is actually gradient boosted decision tree. It increases speed and performance and can be used on a wide range of problems. It controls over-fitting making the model robust. We did regression with squared error. Again 'random_state = 42' is set.

h) Light Gradient Boosting Machine (LightGBM) is a framework that also used tree based learning that produces trees one leaf at a time. It can handle large amount of data. So speed and efficiency increases. The other benefit is lower memory usage. 'random_state = 42' as usual.

i) CatBoost is also a gradient boosting method build on decision tree but specially designed for categorical variables. It usages ordered boosting, a special algorithm for the distribution of data between training and test. It gives state-of-the-art results without complete data training unlike other machine learning network.

j) SVR (Support Vector Machine) supports linear as well as non-linear regression. It fits the error within a certain threshold and handles non-linear data by kernels which uses function like linear, polynomial, radial etc. In our case, we used linear function for kernels.

k) K-Nearest Neighbors, an extension of KNN, predicts the output based on the nearest neighbors values (k). Euclidean, Manhattan etc are used to measure the distance to get the k-nearest neighbors. Finally the model averages the values. The parameter we used 'n_neighbors = 5' which is the number of neighbors for kneighbors queries.

l) Multilayer Perceptron (MLP) Regression is a feedforward artificial neural network (ANN). It has input layer, hidden layer and output layer. It uses activation function (not for input nodes) which is non-linear. MLP minimizes error by modifying weights by backpropagation. We used two hidden layers 'hidden_layer_sizes = (100, 50)', one has 100 neurons; another one has is 50 neurons. 'activation = identity' is used.

m) Deep neural network uses multiple hidden layers. Our code incorporates activation function as relu and optimizer as adam and mean squared error as loss.

The performance of the different machine learning models was evaluated using various methods. Every model takes time to train the model. Training the MLP model took longer than the other models used in this paper. Regarding the prediction time, the Deep Neural Network took more time to predict the output. Outliers of data points mean they are significantly different than most data points in a dataset. Robustness to outliers of all the models was checked in this research. High robustness to outliers indicates performance well even though the outliers are present in the dataset of a particular model. Linear, Ridge, Lasso, ElasticNet, SVR, and Deep Neural Network models didn't show better robustness to outliers compared to others. The scalability of a particular model means the ability to work efficiently with the increase of data without a significant increase in processing time, memory usage, etc. It means how well the model performs if the complexity of the dataset or problem increases. If a model has a high level of scalability, the training time of the model will be less. MLP and CatBoost models take more training time, which means poor scalability. These are the comparisons of the performances of all models. All models are used to find out the water prediction. A person's predicted water required value is then multiplied by the number of family members to calculate the total water needed for a particular family. Suppose the proposed design is used in a house. In that case, the house owner will give input on the water tank capacity, and the Python code will calculate the percentage of water needed from the water tank for all the families in a big house depending on the predicted values calculated before. This concept can be implemented anywhere when proper water distribution and control is a prime issue.

3.4 Arduino Code and Related Circuits

Then, these percentage values will be the input of the Arduino C++ code. Water will be supplied through a water flow sensor. The water flow sensor measures the amount of water that flows through it in liters. Water flows not only through the water flow sensor, it also flows through the solenoid. So, water flows to a family through a water flow sensor and a solenoid. The water flow sensor reads the amount in liters, and the solenoid will ordinarily open first. Maximum water demand is already given as an input, so it will automatically close when the maximum value reaches the solenoid. That is how water will be distributed appropriately, and water wastage can be reduced.

Chapter 4.
SIMULATION

4.1 Google Colab

Google Colab is a cloud-based platform that is provided by Google for writing and running Python codes in a web browser. It is a Jupyter Notebook-like interface coding platform. The Google Colab platform has been used to write and run all the Python codes, machine learning parts, and inputs from the owner and user.

4.2 Arduino IDE

Arduino Integrated Development Environment (IDE) is an open-source software platform for writing and uploading that allows users to code to the Arduino Board. It is also used in the proposed design to implement the microcontroller circuitry.

4.3 Tinkercad

Among the identified software with applications in design and 3D modeling, Tinkercad was seen as useful in simulating as well as validate the running code of a water distribution system. It facilitated the verification of the proper operation of the system before going to the physical model, as various problems and proper functioning would be enhanced in the given model.

Simulation Components and Substitutions:

To emulate the real-world components effectively within the Tinkercad environment, specific substitutions were made. Based on the need to have the real components depiction within the Tinkercad environment some of the swaps were as follows:

Analog Inputs as Water Flow Sensors: Actual water flow sensors were not incorporated in the design but instead analog input pins where incorporated, the pin codes being A0 and A1. These were used as surrogates of the water flow sensors and used to record the simulated flow data. LDRs were employed to mimic these sensors and to change the resistance in responsive to the change in the intensity of the light as compared to that of the water flow.

Motors as Solenoids: In other side, solenoid valves which rely on the opening or the closing of the water passages were used, in this design, two motors were incorporated. These motors resembled the functional utility of solenoids and could be toggled either 'ON' or 'OFF' by the control signals thus emulating the regulation of water supply.

The hardware of the system; the microcontroller of the system Arduino UNO R3; did not undergo any modification as well. During the conducted simulation this microcontroller was programmed with the real code for the physical circuit and thereby the logic as well as the control sequences were exercised as are in real life.

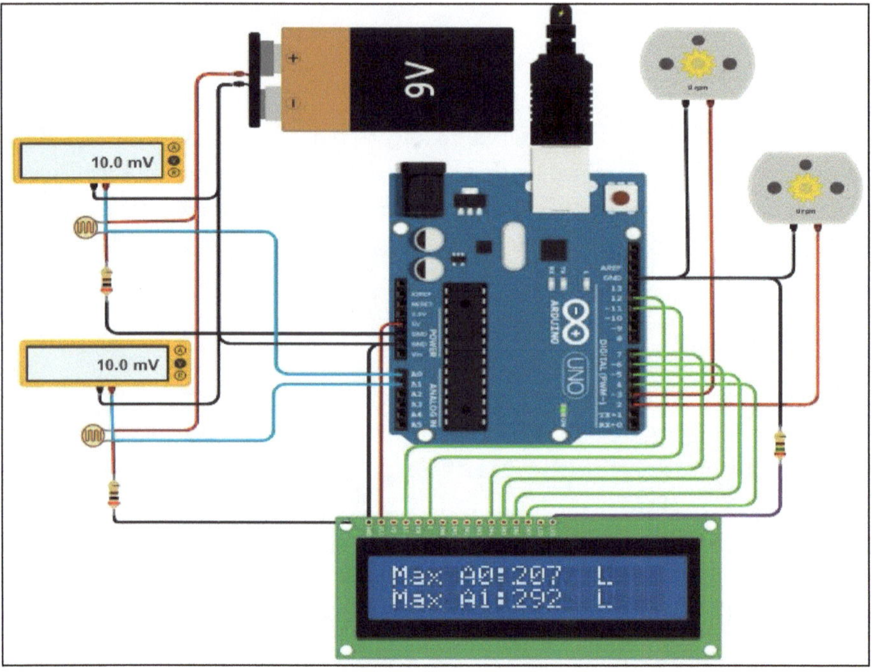

Figure 2: Simulating circuit-1 shows the maximum water demand of each family

Operational Dynamics of the Simulated Circuit:

As depicted in Figure 2, the simulation was structured to allow the owner or operator to input specific water demand values for two families.

Family 1 (A0): Used to a maximum water requirement from the system of 207 liters.

Family 2 (A1): With a higher threshold fixed to 292 liters per cycle.

These inputs set concrete ceilings to the amount of water that each family could use. During the simulation, the analog inputs observed the 'rate' of flow of the water (emulated using the LDRs), the Arduino then measured this data to pump the motors (these served as solenoids).

An integrated LCD display was incorporated to give real-time feedback as feedback that was the quantity of water provided to each family. The working of this system made it possible for the water supply to be active for each of the families until their maximum demand was likely to be exceeded.

System Response and Validation:

Here, the system's responsive behavior is evident:

Family 2: When it reached the maximum allocated water supply of 292 liters, the LCD flashed the word "OFF" and this meant that the water supply to the Family 2 was cut off. Similarly the motor of the solenoid of Family 2, which controls the water flow in the simulation was switched off.

Family 1: Otherwise, when Family 1 had not reached 207 liters, which is the limit, the letters 'ON' were still displayed on the LCD and the concerning motor was on to imitate the open solenoid.

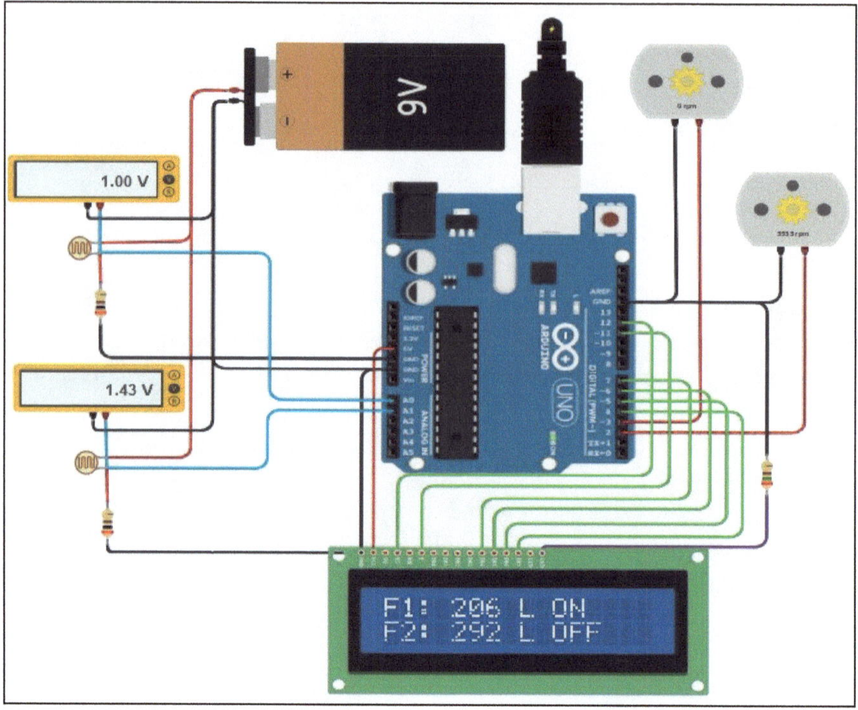

Figure 3: Simulating circuit-2 shows present status

Comprehensive simulation conducted through Tinkercad was very helpful in as far as ensuring that the code as well as circuitry of the water distribution system was functional and would not fail at any one time. As expected, some complexity and/or inefficiencies in delivering, managing, and maintaining were simulated through system and control environment rehearsals to determine problems that may have been overlooked that could cause problems for system implementation. This proactive approach not only facilitated the organizational and co-ordination process of the development but also made certain that when the developed modules are transferred from the simulative environment to the real, they would be as effective.

Chapter 5.
PRACTICAL IMPLEMENTATION

The circuit is also practically implemented. As the simulation worked perfectly, the practical implementation was expected to work as the proposed model. Practical circuits were set up to supply all the families through a water flow meter and solenoid, and all other processes were built according to the proposed design. There was an LCD for every user or family to show the water supply amount to that particular family. The implementation was successful, and all the circuitry parts worked well according to the design.

5.1 Arduino Code: Sample Arduino code is given below:

```
#include <LiquidCrystal.h>

int numInputs;
// Number of analog inputs specified by the user
int *maxAnalogValues;
// Maximum analog values for each input
LiquidCrystal lcd(12, 11, 7, 6, 5, 4);
// Initialize the LCD

void setup() {
  // Initialize serial communication
  Serial.begin(9600);
```

```
// Initialize LCD
lcd.begin(16, 2);

// Get the number of inputs from the user
lcd.print("Total Inputs:");
Serial.println("Enter the number of analog inputs: ");
while (!Serial.available()) {}
numInputs = Serial.parseInt();
Serial.println(numInputs);
// Display the entered number of analog inputs
lcd.setCursor(13, 0);
lcd.print(numInputs);
delay(2000);
// Dynamically allocate memory for maxAnalogValues array
maxAnalogValues = new int[numInputs];

// Get the maximum values for the analog inputs from the user
lcd.clear();
for (int i = 0; i < numInputs; i++) {
  lcd.setCursor(0, i);
  lcd.print("Max A");
  lcd.print(i);
```

```
  lcd.print(":");
  Serial.print("Enter the maximum value for Analog Input ");
  Serial.print(i);
  Serial.println(": ");
  while (!Serial.available()) {}
 // Wait for user input
  maxAnalogValues[i] = Serial.parseInt();
  Serial.println(maxAnalogValues[i]);
  lcd.print(maxAnalogValues[i]);
  lcd.print("  L");
 }
 delay(2000);
 // Initialize analog inputs and digital outputs based on user input
 for (int i = 0; i < numInputs; i++) {
  pinMode(A0 + i, INPUT);
 }
 for (int i = 0; i < numInputs; i++) {
  pinMode(2 + i, OUTPUT);
  digitalWrite(2 + i, HIGH);
 // Initially turn on all outputs
 }
}
```

```
void loop() {
 int *analogReadings = new int[numInputs];
 // Current analog readings
 char outputState[3] = "ON";  // Digital output state

 // Read analog values
 for (int i = 0; i < numInputs; i++) {
  analogReadings[i] = analogRead(A0 + i);
  Serial.print("Analog Input ");
  Serial.print(i);
  Serial.print(": ");
  Serial.println(analogReadings[i]);
 }

 // Display all analog readings and output states on LCD
 lcd.clear();
 for (int i = 0; i < numInputs; i++) {
  lcd.setCursor(0, i);
  lcd.print("F");
  lcd.print(i+1);
  lcd.print(":");
  lcd.setCursor(4, i);
```

```
lcd.print(analogReadings[i]);

lcd.setCursor(8, i);

lcd.print("L");

// Check if analog value exceeds maximum

if (analogReadings[i] >= maxAnalogValues[i]) {

  digitalWrite(2 + i, LOW);

// Turn off digital output

  outputState[0] = 'O';

  outputState[1] = 'F';

  outputState[2] = 'F';

  outputState[3] = ' ';

  outputState[4] = ' ';


} else {

  digitalWrite(2 + i, HIGH);

// Turn on digital output

  outputState[0] = 'O';

  outputState[1] = 'N';

  outputState[2] = ' ';

  outputState[3] = ' ';

  outputState[4] = ' ';

}
```

```
  lcd.setCursor(10, i);

  lcd.print(outputState);

 }


 delete[] analogReadings;

 // Free memory

 delay(1000);

 // Delay for 1 second

}


void cleanup() {

 delete[] maxAnalogValues;

 // Free memory

}
```

This arduino code is after the machine learning model predicts the amount of water needed by each family. The code is a sample code. This layout of arduino is used for multi input analogue and controlling the digital output regarding the maximum threshold initialize by the user. Upon initializing, the program waits for the user to input over the serial monitor the number of analog inputs and the respective maximum values the reading might assume. It is only at this stage that such values are displayed in an LCD screen. In this mode the system repeatedly scans the specified analog inputs, for example from sensors and compares them to the maximum. As an input value crosses its maximum value corresponding to it, the Digital output like a motor or solenoid is switched off and on the LCD “OFF” will be displayed, whereas, if its value is less than this threshold, the Digital output remains on, the LCD will display “ON” For this, there is dynamic memory allocation to store the maximum values and for cleaning up; they release the memory. The overall functionality enables one to monitor several inputs and control several outputs in real time and this makes it suitable in application such as control of watering systems.

The practical demo implementation is shown in figure-4. It can be easily implemented in residential building or in a metropolitan area.

5.2 Implementation:

Water Tank

The 'highest blaze' of this system is a very large and obvious green water container, which serves as a tank for the entire system. This is the principal supply of water within the compound and feeds the distribution system which delivers water to each distinct family unit. The placement of the tank in the system is also important because they help provide water to the system hence help the system operates as required.

Figure 4: Practical implementation of the circuit

Piping System

A series of green pipes carry water from the central tank to other portions of the system but not along the circle outer edge. In every section a different family is displayed and lines to convey water in the right manner to the taps is linked to solenoids/motorized valves. The manner in which the water is piped is very systematic and the layout is of immense importance in regulating the path and quantities of water supplied to every family.

Flow Meters and Solenoids

All the sectional water supply must incorporate a flowmeter in the inlet of the family section and surrounding water supply sections. These flow meters can be best described as being part of the 'system' and they help measure the flow rate of water with the help of a display which relays the information to an Arduino microcontroller. In this design the solenoid valve only or motors used instead of solenoid valves in the previous type control the water flow. These valves control water flow in the reactors with aid of flow meters but all opened or closed by Arduino thus the water flow is controlled and is not supplied beyond the capacity set.

LCD Displays

A LCD display is used in each of the family sections for expressing the state of operation of the water distribution system and are fixtures which are critical for substantiating the feedback mode. This kind of displays is used to relay all sort of information for instance the amount of water that has been availed to each family. In this case the present flow can be divided by the maximal allowed flow so that the user is in a position to know if the water supply is still on or off. The LCD displays therefore functions as the interface that the user will use to monitor and control the distribution of water.

Arduino Microcontroller and Relay Modules

At the base level of the system there are two Arduino microcontrollers that constitute the raw mechanics of the system. Such microcontrollers are in actuality the ones that 'execute' through the code which controls each and every process of the water distribution system. The Arduino receives data from the flow meters and after calculation of a number of factors gives signals to the relay modules. The relay modules also act as an interface from the low power Arduino system to the right power required to make solenoid valves or motors ON. In particular, it enhances the possibility of offering optimum functional deployment catering to the inputs and conditions offered by the user.

Wiring and Power Supply

The wiring that has been done to connect the different components of the system is well done to ensure that the sensors, Arduino microcontrollers, and actuators interface well. This well defined wiring structure is used to keep check on the authenticity of the signals which are coming from the flow meters and passing through the arduino box and then to the relays and valves from the arduino box. External source of power is used and this powers all the devices including the Arduino, relay modules and other in order to enable them to perform as desired.

Functionality of the System

It comprises of the water supply to separate families and the flow of the water to the families is regulated for a fixed consumptive rate. The flow meters in turn continuously measure the water used by each family and feed back their information to the Arduino. If the consumption of water by a family has reached the maximum permissible value, the corresponding solenoid valve is turned off and the water does not go. This is followed on the LCD display by an 'OFF' message in cases where the supply has been turned off. On the other hand, should the water flow be below the required limit, the solenoid remains open and will allow the water to pass through and the ignite button light display will reflect "ON", this is to convey that the supply is still on.

This is a well-coordinated model that is best illustrated in the water distribution system where resources are also well utilized within a practical system. This way, the Arduino microcontroller in monitoring and controlling the supply of water reduces wastage of water while at the same time only providing every family with the right amount of water needed. This success is another proof of the efficiency of the first phase, the representation of the simulation phase which helps implement the final system and gain a tactile model of it without the alteration of the model on the real terrain.

Chapter 6.

THE RESULT, MODEL PERFORMANCE, AND DISCUSSION

6.1 Predicted Result

The images show process and result of a water distribution system where an open machine learning model estimates the needs of water for families according to the inputs given by the user.

```
Enter total number of families: 2
Enter the water tank capacity in liters: 500
```

Figure 5: Owner Input (sample values)

In Figure 5, the owner set-up the number of families living in a house, it contains 2 families (sample) and capacity of water tank is 500 liters. This is important as it sets the wag or rate which water is distributed at.

```
Enter your family number: 2
Enter total number of family members for Family 2: 3

Family 2 details:
Enter Religion (islam, christianity, hinduism, buddhism, others): islam
Enter Location (urban, rural): urban
Enter Region (North America, Europe, Asia, Africa, Middle East, Australia, South America): australia
Enter Occupation (office, labor, student, jobless): student
Enter Temperature (hot, moderate, cold): cold
Enter Season (winter, spring, summer, fall): fall
```

Figure 6: User-1 Input (sample values)

Figures 6 and 7 present inputs in the precipitation of each family in a detailed manner. In Figure 6 the owner specifies characteristics of the second family such as the number of people (3), religion (Islam), place of living (urban), country (Australia), job (student), temperature (cold), and season (fall).

Likewise, in Figure 7: Family 1 inputs are defined, where the number of family members is 2, religion (others), place of residence (rural), region (Asia), occupation (labor), temperature conditions (hot), season (summer). It is crucial to mention these details because the distribution of water has to be calculated with the help of machine learning model according to certain environmental conditions and the way of life.

```
Enter your family number: 1
Enter total number of family members for Family 1: 2

Family 1 details:
Enter Religion (islam, christianity, hinduism, buddhism, others): others
Enter Location (urban, rural): rural
Enter Region (North America, Europe, Asia, Africa, Middle East, Australia, South America): asia
Enter Occupation (office, labor, student, jobless): labor
Enter Temperature (hot, moderate, cold): hot
Enter Season (winter, spring, summer, fall): summer
```

Figure 7: User-2 Input (sample values)

Figure 8 shows the predicted output where the model partitions water in accordance to the inputs. The output shows that Family 2 gets 292. 94 liters (58.59%) and Family 1 disqualified receiving 207. 06 liters (41.41%). The following is the prediction based on the aforesaid factors showing how the water is distributed in an efficient and effective manner to the families and the amount of wastage minimized is done as under: From the evaluation process using machine learning for water requirement determining based on multiple factors, it can be seen how technology can contribute to rational materials usage in smart homes.

```
Water Distribution:
Family 2: 292.94 liters (58.59%)
Family 1: 207.06 liters (41.41%)
```

Figure 8: Predicted Water Values in Liters

The bar chart of the figure 9 is titled "Feature Importance" and it shows the degree of significance of factors that define level of water usage. The features on the abscissa are; religion: Christianism, Hinduism, Islam others, place of living: urban, region: Asia, Australia, Europe, middle east, north America, south America occupation: labor, office, student temperature: hot, moderate, season: Spring, Summer winter. The y-axis is a measure of these features and the values along it can be seen as the 'importance' of each of them.

But one of the impressive outcomes of this step is that several geographic locations are revealed by the fitting as significantly important; thus indicating that the geographic location of regions does matter for the water demand prediction. Prominent value is found to be highest for Middle Eastern nations, value nearly to 100 and for Europe and North America represented value is important but ranked lower than Middle East nations. But other than religion, majority of the religions represent a weak to no influence on water demand as illustrated in the figure below. Of all the features, thus, the "Region_Middle East" has the highest importance indicating that regionality appears to have a very strong impact on the water demand. Only variables linked to occupation types, namely types of offices and clerical occupations as well as temperature variables are of medium importance only to the model.

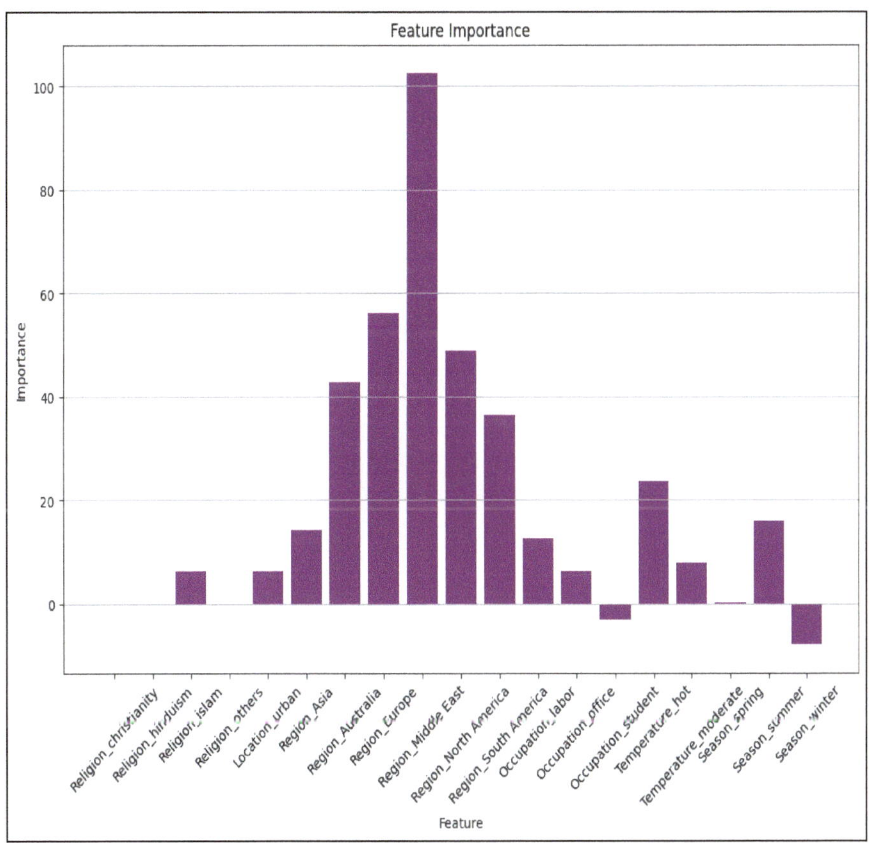

Figure 9: Feature Importance

Rather bewilderingly, those few features below indicate negative importance, and this might indicate certain issues with the given data set as well as the structure of the model that was built. For example, ''Season_Winter'' has a negative importance which makes the variable a negative impact, or a problem to the model. There is then a rather minor issue with the proposed design – that may need further inspection and potentially the data or the modeling so as to achieve sufficient prediction power.

Altogether, the chart reiterates that geographic/ regional factors are highly influential in determining water demand; the religious factors, apart from the Islam factors, more or less important. The aspect of the features' contribution that is as delicate as this one is invaluable both when constructing sound models and analyzing the technologies deployed in the sphere of water supply to make efficient decisions.

6.2 Different Machine Learning Models Performance

Many machine learning regression models, such as linear, Ridge, Lasso, ElasticNet, Decision Tree, Random Forest, XGBoost, LightGBM, CatBoost, SVR, and KNN, are used to build up the concept of this paper. MLP and deep neural networks are also used to measure prediction performances. Many factors were considered to evaluate the models' performance. Table 2 shows the MSE values of the validation and testing data.

Table 2. *MSE Comparison of Different Models*

Sl. No.	Model	Validation MSE	Testing MSE
01	Linear	1.17e-26	1.14e-26
02	Ridge	0.80	0.65
03	Lasso	1.26	1.01
04	ElasticNet	19.78	15.98
05	Decision Tree	214.35	250.34
06	Random Forest	144.88	128.37
07	XGBoost	29.69	27.39
08	LightGBM	34.53	25.78
09	CatBoost	15.99	13.48
10	SVR	0.01	0.01
11	KNN	1575.74	1889.02
12	MLP	0.01	0.01
13	Deep Learning	197.24	172.56

We have tabulated the result of the generalization capability of various machine learning models quantifying them in terms of MSE for the validation set and test set.

For all the various models, as with the MAE value the MSE value has the lower the better scenario which implies that the closer the predicted values the better the performance of the model. Linear regression has delivered the lowest MSE for both validation and testing that was nearly zero showing that the model was nearly perfect in predicting the results. Similarly, other investigated models like Ridge, Lasso, SVR and MLP achieved very low values of MSE suggesting a very good predictive accuracy.

As we can see in the case of ElasticNet, which is the combination of the penalties of both the methods namely Ridge and Lasso, the MSE values though are slightly higher one more time indicates that though the prediction accuracy might not be very high should be reasonably acceptable. Among other tree-based models, Decision Tree and Random Forest yield higher MSE's, especially Decision Tree has the highest one, which brings a couple of concerns as to the applicability of such models together with the potential of the models learning the training data so well. However, all kinds of gradient boosting techniques for example XGBoost and LightGBM offer improved performance but with slightly higher MSE which precisely proves overall practical ability of effective forecasting; and at slightly lower MSE value CatBoost also do the similar task.

As expected, setting up very little values for one's neighbors, and more abruptly, the K-Nearest Neighbors (KNN) yields significantly high MSE values and is therefore annulled for this dataset. Similar to the MLP method, Deep Learning methods have higher MSE values and therefore it can be said again that they may overfit and or need more fine tuning since as is presented these are more complex models than SVR. In general, and in accordance with table 2, simple linear models and SVR are very accurate models, but so are all the complex models like all the ensemble methods and neural networks, at the cost of sensitivity to tuning and other factors such as training time, prediction time and sensitivity to outliers and scalability. They are necessary parameters determining, which model could be adopted for the practical purposes in the future.

Figure 10 – Training Time

In figure 10 the training times for different models has been depicted. Out of all the evaluated models, the MLP (Multilayer Perceptron) Regressor training is seen to take almost 6 seconds, which is noticeably longer than the others. Both the CatBoost and LightGBM regressors have relatively long training time albeit much shorter than that of the MLP Regressor. The next models are Deep Neural Network, SVR, KNN, XGBoost, Random Forest, Decision Tree, ElasticNet, Lasso, Ridge, and Linear

Regression; and all of them show very small training times suggesting that all of them have efficient training.

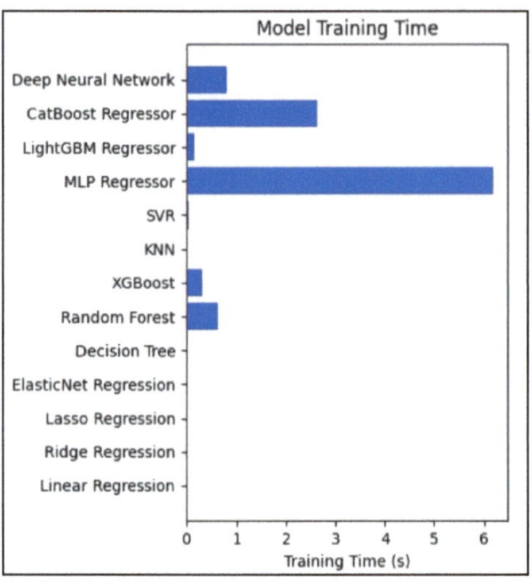

Figure 10: Training Time of the Models

Figure 11: Time of Model Prediction

Figure 11 illustrates the prediction times of the models analyzed in this study. The Long Short Term Memory model ranks the lowest in terms of the time it takes to make the predictions at nearly 0. With the internal activation time of 20 seconds, which considerably exceeds other models. This points out the fact that, the Deep Neural Network can produce good performance, though it takes some time for the network to make the prediction. According to the above table the SVR model also indicates predictive time slightly more than the other models but significantly less than DNN. As for other models, CatBoost, LightGBM, MLP, KNN, XGB, RF, DT, ElasticNet, Lasso, Ridge, and Linear Regression models have almost equal and very low prediction time suggesting pretty fast prediction.

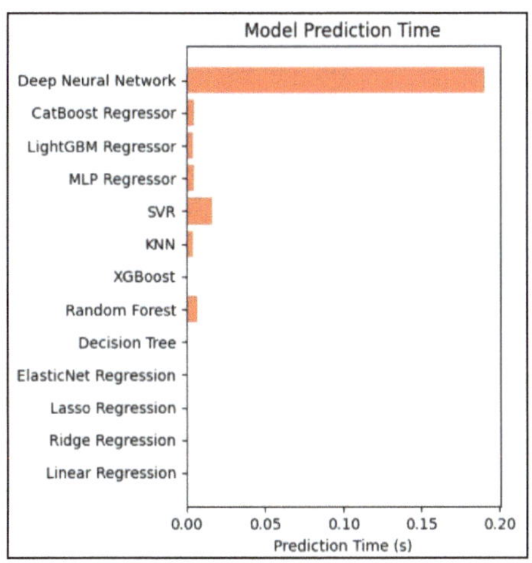

Figure 11: Prediction Time of the Models

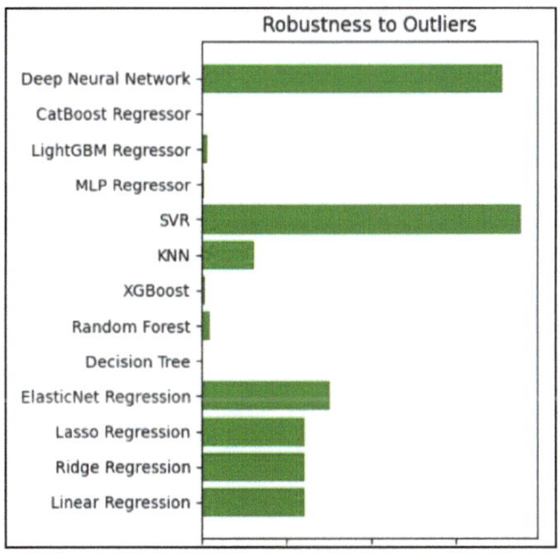

Figure 12: Robustness to Outliers of the Models

Figure 12: tolerance to outliers

Figure 12 shows the comparison of the models in terms of their resistivity to outliers. Robustness is better with a short bar. From the above bar chart, it is evident that all the three models, which include Deep Neural Network, SVR, and ElasticNet Regression models have very short bars and thus showing high robust to outliers meaning the models are insensitive to the presence of outliers in the data set. For CatBoost, LightGBM, MLP, Random Forest robustness is also moderate but to a lesser degree. Lastly, there is an (other) interesting point compared to the other models, KNN, Decision Tree, Lasso, Ridge, Linear Regression models seem to be less robust as they have longer bars, which might mean that attempts to utilize the data in some malicious way may influence the model more than others.

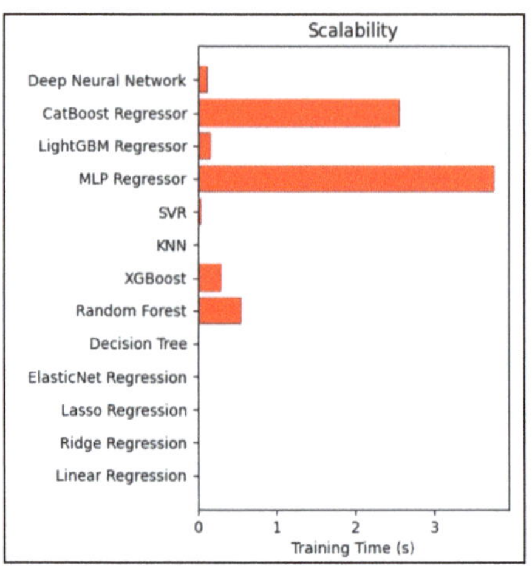

Figure 13: Scalability of different models

Figure 13: Scalability

As depicted in Figure 13 below, all the models were able to grow in their ability to process more information as and when it was fed to them. Deep Neural Network, SVR, KNN, XGBoost, Random Forest, Decision Tree, Elastic Net Regression, Lasso Regression, Ridge Regression, and Linear Regression have been found to perform relatively fast with slight increase in training time as the size of data increases. But, the MLP Regressor and CatBoost Regressor have relatively high-level training time, which show the signal of scalability problems. This implies that although these models may be effective if the number of training examples is small, the effectively of the models may reduce as the number of training examples added increases.

So, the Decision Tree and LightGBM models are the best for this work compared with the other models by analyzing all performance factors.

CONCLUSION

In this study, it is investigated that the wastage of water, water distribution, and management are some big problems globally. This proposed design tried to provide a better solution to the listed issues. The result of the proposed design shows the reliability of the work, and it can be applied effectively and practically. Water usage prediction based on the factors in this paper is a new concept, and the distribution system is also a new method for this research. The machine learning prediction, water percentage calculation according to the various inputs, and finally, the practical application of the project based on the Arduino microcontroller all worked adequately. The performances of different machine learning models were also experimented with, and they were very consistent for a few models, while some models' performances were poor. But still, in the feature importance graph, we see some issues of negative values, which should be solved in further study. In conclusion, the proposed method in this paper provides a valuable contribution to the related field and highlights the importance of further research in this area.

Future Work

It is described that this proposed work provides a valuable contribution to the related field. However, it is necessary to address some of the job limitations so that future research can be done in this area. The dataset should be prepared with more data from all over the world, and as many factors as possible should be considered to make the predicted value more accurate and perfect. A water flow meter sensor is used in this work, so billing according to water usage can be easily added to the project. The design in this paper was mainly focused on the house and family areas, but it can be further implemented even in the whole country using the same concept so that further study can make this research ahead. This topic is still a problem to be solved, so the future may bring new additions to the topic that will help to manage, distribute, and control water, reduce this compulsory element of nature for our lives, and make our lives safer.

References

Adu-Manu, K. S., Adjetey, C., &Apea, N. Y. O. (2022).Leakage detection and automatic billing in water distribution systems using smart sensors. In *Digital Transformation for Sustainability: ICT-supported Environmental Socio-economic Development* (pp. 251-270). Cham: Springer International Publishing.

Ali, A., Rahman, M., Siddique, M., Galib, S., &Nashiry, A. (2023).Design of an automatic rooftop water tank filling system and measurement of consumed water for home appliance.*International Journal of Automation and Smart Technology*, *13*(1), 2371-2371.

Ali, S., Gillani, S. M. D., Khan, S. A., &Zain, S. M. (2024). Controlled fluid flow without controlling pump using arduino.

Almulhim, A. I., &Abubakar, I. R. (2024).A segmentation approach to understanding water consumption behavioral patterns among households in Saudi Arabia for a sustainable future.*Resources, Environment and Sustainability*, *15*, 100144.

Babu, P., &Rajasekaran, C. (2016).Economically precise water resource management for domestic usage in India.*Circuits and Systems*, *7*(10), 2821.

Baranidharan, S., Hariprasad, T., Nanthakumar, C., & Kumar, C. (2023, September).Smart water distribution system using internet of things (IoT).In *AIP Conference Proceedings* (Vol. 2831, No. 1).AIP Publishing.

Bergel, T., &Młyńska, A. (2021).Analysis of the impact of the air temperature on water consumption for household purposes in rural households.*Journal of Ecological Engineering*, *22*(3), 289-302.

Chang, H., Praskievicz, S., &Parandvash, H. (2014). Sensitivity of urban water consumption to weather and climate variability at multiple temporal scales: The case of Portland, Oregon. *International Journal of Geospatial and Environmental Research*, *1*(1), 7.

Chavhan, S., Gawande, U. H., Bhoyar, D., Rangari, S., Gardi, M., &Gaikwad, A. (2024).IoT-based remote control water distribution system.*International Journal of Intelligent Systems and Applications in Engineering*, *12*(7s), 572-583.

Crouch, M. L., Jacobs, H. E., & Speight, V. L. (2021). Defining domestic water consumption based on personal water use activities. *AQUA—Water Infrastructure, Ecosystems and Society*, *70*(7), 1002-1011.

Dimkić, D. (2020). Temperature impact on drinking water consumption.*Environmental Sciences Proceedings*, *2*(1), 31.

Fan, L., Liu, G., Wang, F., Geissen, V., &Ritsema, C. J. (2013). Factors affecting domestic water consumption in rural households upon access to improved water supply: Insights from the Wei River Basin, China. *PloSOne*, *8*(8), e71977.

Gore, S., Dutt, I., Dahake, R. P., Khodke, H. E., Kurkute, S. L., Dange, B. J., & Gore, S. (2023). Innovations in smart city water supply systems. *International Journal of Intelligent Systems and Applications in Engineering*, *11*(9s), 277-281.

Kalyani, C., Pradnya, K., Samkit, C., &Patil, V. (2023).Regular and equal water supply system.*SAMRIDDHI: A Journal of Physical Sciences, Engineering and Technology*, *15*(01), 34-37.

Kumar, B. S., Ramalingam, S., Balamurugan, S., Soumiya, S., &Yogeswari, S. (2022, October). Water management and control systems for smart city using IoT and artificial intelligence. In *2022 International Conference on Edge Computing and Applications (ICECAA)* (pp. 653-657). IEEE.

Matos, C., Briga-Sá, A., Pereira, S., & Silva-Afonso, A. (2013, September).Water and energy consumption in urban and rural households.In *Proceedings of the 39th International Symposium CIB W062 on Water Supply and Drainage for Buildings, Nagano, Japan* (pp. 17-20).

Nirmala, D., Pooja, G., Sowmya, U., Mohamed, A., Durdona, A., Rajavarman, R., &Parkunam, N. (2023).System for water quality monitoring and distribution.In *E3S Web of Conferences* (Vol. 399, p. 01016).EDP Sciences.

Nur'Im, J., &Setyawan, G. (2023, January).Prototype design of clean water distribution system on residential scale using Arduino Mega microcontroller.In *AIP Conference Proceedings* (Vol. 2540, No. 1).AIP Publishing.

Ogidan, O. K., Olla, M., &Odey, I. (2022, April). Water distribution control with real-time monitoring.In *2022 IEEE Nigeria 4th International Conference on Disruptive Technologies for Sustainable Development (NIGERCON)* (pp. 1-5).IEEE.

Pratama, M., &Firmansyah, M. (2024).AWAS (Automatic Water System) as a clean water distribution device to mitigate the impact of climate change.

Rahmatulloh, A., Supriatna, G. T., Widiyasono, N., &Darmawan, I. (2023).IoT-Enabled water distribution monitoring: a sensor-based analytical model.*IngénierieDes SystèmesD'information*, *28*(6).

Reddy, D. G., V, D., Salanke, N. G. R., & MN, M. (2024).Aadhaarenabled water distribution system.*Water Resources Management*, 1-13.

Rexline, S. J., Renold, M., &Deeba, M. (2023, October).Potable water distribution monitoring system using internet of things.In *AIP Conference Proceedings* (Vol. 2842, No. 1).AIP Publishing.

Rodríguez, C., Sánchez, R., Lozano-Parra, J., Rebolledo, N., Schneider, N., Serrano, J., &Leiva, E. (2020). Water balance assessment in schools and households of rural areas of Coquimbo region, north-central Chile: Potential for greywater reuse. *Water*, *12*(10), 2915.

Rondinel-Oviedo, D. R., & Sarmiento-Pastor, J. M. (2020). Water: consumption, usage patterns, and residential infrastructure. A comparative analysis of three regions in the Lima metropolitan area.*Water International*, *45*(7-8), 824-846.

Singha, B., Eljamal, O., &Karmaker, S. C. (2024).Changing patterns of household water consumption and conservation behaviour in Bangladesh: an exploration in the context of COVID-19 pandemic.*International Journal of Innovation and Sustainable Development*, *18*(1-2), 106-122.

Smith, A., & Ali, M. (2006).Understanding the impact of cultural and religious water use.*Water and Environment Journal*, *20*(4), 203-209.

Tejas, T., Yadav, U., Krishnan, V., Vishrutha, V., &Mallikarjuna, M. (2024, February).IoT based water management system with machine learning. In *AIP Conference Proceedings* (Vol. 2742, No. 1).AIP Publishing.

Timmerman, M. (2013). Water consumption in the Middle East and North Africa (MENA) region: a virtual water approach.

Utami, R. R., Geerling, G. W., Salami, I. R., Notodarmojo, S., & Ragas, A. M. (2024).Mapping domestic water use to quantify water-demand and water-related contaminant exposure in a peri-urban community, Indonesia.*International Journal of Environmental Health Research*, *34*(1), 625-638.

Velayudhan, N. K., Pradeep, P., Rao, S. N., Devidas, A. R., & Ramesh, M. V. (2022). IoT-enabled water distribution systems—A comparative technological review.*IEEE Access*, *10*, 101042-101070.

Vladislav, D. S. (2022).A short study on smart water distribution through sensor network.*Big Data and Computing Visions*, *2*(3), 122-127.

Xenochristou, M., &Blokker, M. (2018, July). Investigating the influence of weather on water consumption: a Dutch Case Study:(032). In *WDSA/CCWI Joint Conference Proceedings* (Vol. 1).

Yan, L. (2015). *The ethnic and cultural correlates of water consumption in a pluralistic social context–the Sydney Metropolitan Area* (Doctoral dissertation).

www.ingramcontent.com/pod-product-compliance
Lightning Source LLC
LaVergne TN
LVHW021306160826
845679LV00001B/226

* 9 7 9 8 8 9 2 4 8 5 8 4 5 *